Beef Cattle and Swine
Growing Beef, Fattening Range Steers in the Winter and Feeding Pigs

by Minnesota Agricultural Experiment Station

with an introduction by Jackson Chambers

This work contains material that was originally published in 1898.

This publication is within the Public Domain.

*This edition is reprinted for educational purposes
and in accordance with all applicable Federal Laws.*

Introduction Copyright 2018 by Jackson Chambers

Self Reliance Books

Get more historic titles on animal and stock breeding, gardening and old fashioned skills by visiting us at:

http://selfreliancebooks.blogspot.com/

Introduction

I am pleased to present another title in the "Raising Pigs" series..

As with all reprinted books of this age that are intended to perfectly reproduce the original edition, considerable pains and effort had to be undertaken to correct fading and sometimes outright damage to existing proofs of this title. At times, this task is quite monumental, requiring an almost total "rebuilding" of some pages from digital proofs of multiple copies. Despite this, imperfections still sometimes exist in the final proof and may detract from the visual appearance of the text.

I hope you enjoy reading this book as much as I enjoyed re-publishing and making it available to fanciers again.

With Regards,

Jackson Chambers

BEEF CATTLE AND SWINE.

THOMAS SHAW.

This bulletin contains three experiments. The first of these relates to growing beef, the second to the fattening of range steers and the third to growing pork.

GROWING BEEF IN MINNESOTA.

SECTION NO. 1.

The opinion has heretofore been held by many in our State that beef cannot be grown at a profit on the average farm of Minnesota. To so great an extent has this view prevailed, that it has doubtless virtually hindered numbers from making the attempt to grow it. This opinion was largely grounded on the fact that farther west on the ranges the pastures are free, and that therefore it was an impossible thing to grow beef on an arable farm in which considerable capital had been invested, in competition with beef grown on free pasture. To so great an extent has this opinion prevailed, that it has even been voiced on the public platforms by the teachers of the people. It was not unnatural, therefore, that those who tilled the soil should settle down to the conviction that beef could not be grown at a profit in our State. The writer never sympathized with such a view, and this experiment was undertaken to ascertain, if possible, which was the correct view.

Plan of the Experiment.—In planning the experiment it was proposed to take two calves at birth, to feed them freely on suitable farm products, under the same conditions until nearly two years old, and then to push them to a finish so that they should be put upon the market at an age not beyond 30 months. The food fed was to be such as any farmer could grow, and the care and management were to be of a character that would be easy of imitation by any farmer or by any farmer's boy. A larger number of animals would have been reared thus could suitable representatives have been obtained at the time.

Autumn Calves Chosen.—It was determined to choose autumn rather than spring calves and for the following reasons: 1. When the calves are dropped in the fall the cows are in milk at a season when the milk product is relatively more valuable than in the summer; 2, calves coming in the autumn can be better cared for during the milk period, that is to say, during the most critical period of growth in the life of the calf, for the grower is not pressed with labor at that season of the year as he is in the summer; 3, the calves are in fine condition to go out on the grass as soon as it is ready in the spring, since before the arrival of the grass they are already weaned, and food so palatable should keep them growing right along in good form; 4, they may enter the following winter in fine shape and can, in consequence, be wintered at less outlay relatively than calves dropped in the spring; and, 5, they are at a suitable age for being marketed in the late winter or early spring when meat is relatively dear.

It was also decided that calves should be chosen, which, in breeding would be the offspring of grade cows of good milking qualities, that is to say, the calves of dual purpose cows, sired by a purely bred beef bull. It was not easy to get such calves, owing to the extent to which dairy bulls prevailed in the neighborhood, and indeed in nearly all parts of the state. Mr. H. F. Brown, of Minneapolis, at length came to the rescue, and presented us with a pair of steer calves in the autumn of 1895. Both of these were out of dams that in breeding were essentially grade Shorthorns. One had a dash of Jersey blood, and both were good milkers.

The respective sires were pure Shorthorn bulls belonging to Mr. Brown's herd.

Description of the Animals.—The two steers were named "Jack" and "Prince" respectively. The former was dropped August 1st, 1895, and the latter September 28th following. Both were essentially of the parallelogramie outline, and both were possessed, but not quite equally, of what may be termed the essential features of the beef form in at least a fair degree. Both were a little leggy as will be apparent from the accompanying sketches. Neither was sufficiently blocky for making the highest type of the beef steer, otherwise Jack had what might be termed good all round development. Prince was somewhat inferior to Jack in width and in roundness of spring of rib. The ribs of the latter bore some testimony to the Jersey blood in the ancestry on the dam's side. They had a medium downward pitch, were a little open spaced, especially toward the rear, and were more plainly apparent to the eye when viewed from the side. As is generally known, those features in rib development are not in keeping with the very highest type of beef production. And just here it may be mentioned that the disadvantage at which this steer was placed because of his less perfect conformation did not seem to militate against his growth. It was more especially during the forcing period at the finish that he lost ground, and on the block he was not, in some respects, quite the equal of the other. In other words it would seem to be true, that want of highest excellence in beef type does not hinder development so much during the growing period as during the finishing period, and that the lack of ability in such animals to make pounds of gain is of less consequence to the grower than the lack of quality in the meat, which compels its sale at a lower price.

Foods Fed.—The foods fed were new milk, skim milk, meal, hay, soiling food, ensilage, pasture and field roots. The skim milk was obtained from the dairy and fed while it was yet warm, that is to say, soon after it had been

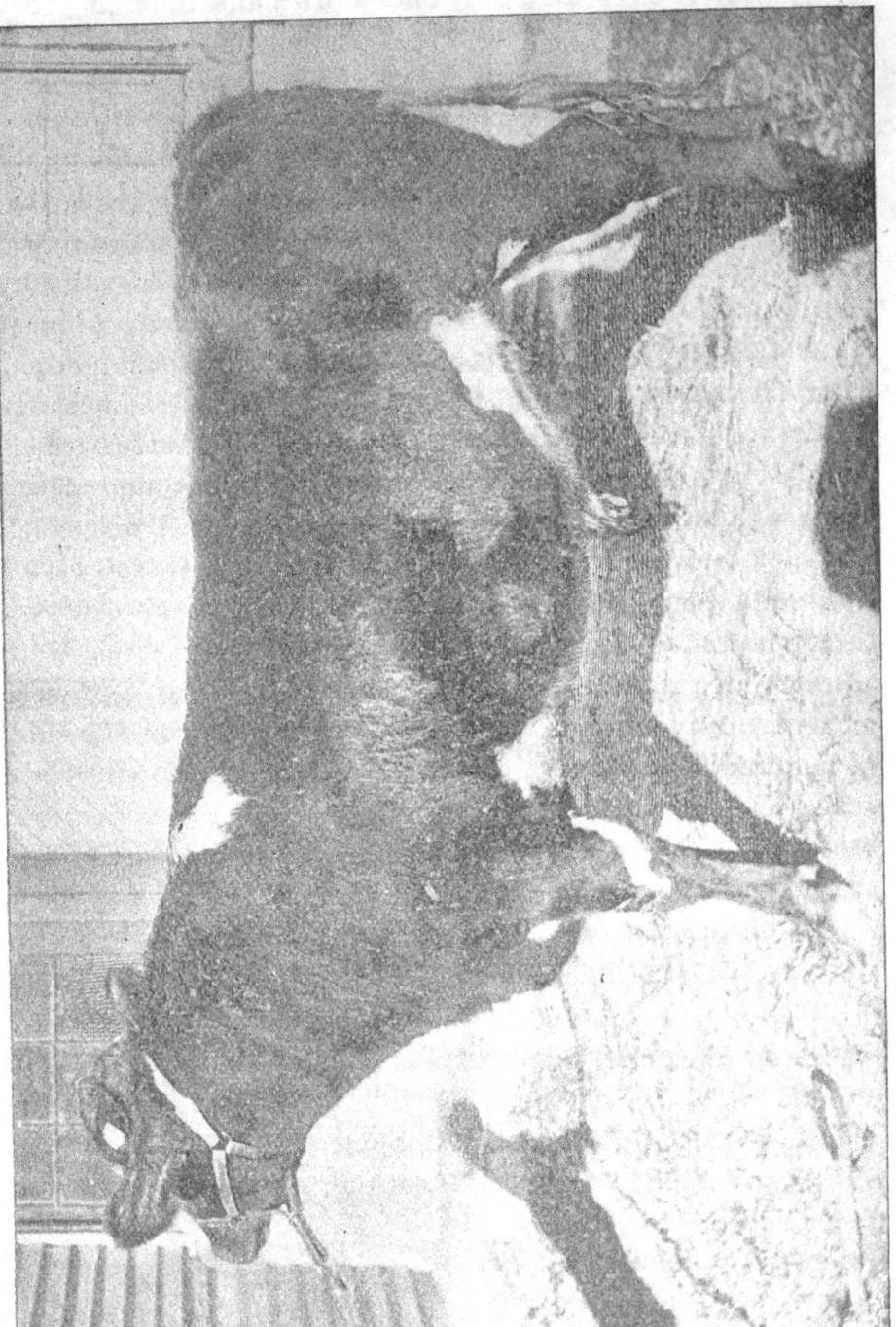

Fig. 1. Jack When Ready for the Block.

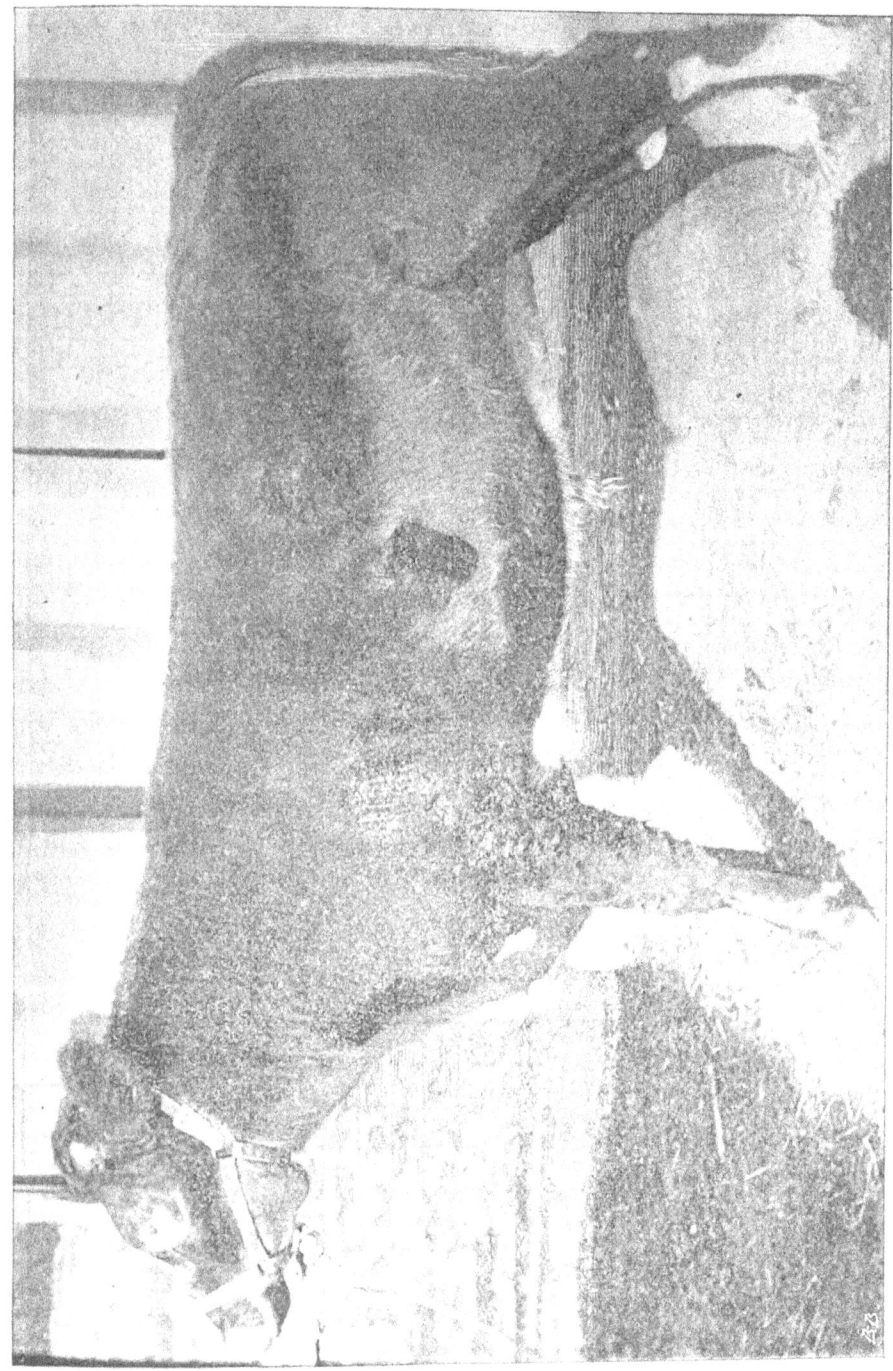

Fig. 2. Prince When Ready for the Block.

separated from the cream. The meal fed included wheat, bran, oats, barley, corn and oilcake. These foods were of course not all fed simultaneously, but in such combinations as were thought prudent from time to time. The grain was all ground. The hay was clover and timothy, chiefly the former, and was fairly good in character until the last winter, when it was chiefly timothy and of poor quality. The soiling food consisted mainly of green corn and green sorghum. The ensilage was corn. The pasture was blue grass, and was plentiful in supply. The field roots consisted mainly of mangels.

Food Values.—The food fed was charged at market values. These values represent the average price in the State that the farmer would obtain for such products when delivered at the ordinary place of sale, so far as those could be ascertained. The bran and oilcake were valued at the average cost to the farmer delivered on the farm, since these have usually to be purchased by the feeder. The prices for the foods fed are given below, and as they necessarily vary from year to year, the variations are also given. The following were the values for the last six months of 1895 and the first six months of 1896:

New milk per 100 pounds	$0.63
Skim milk per 100 pounds	.12
Wheat bran per ton	6.50
Oats per bushel of 32 pounds	.14
Corn per bushel of 56 pounds	.18
Oilcake per ton	14.00
Clover hay per ton	3.50
Green food per ton	.75
Mangels per bushel of 50 pounds	.04½

During the last six months of 1896 and the first six months of 1897, the barley fed was charged at 16 cents per bushel of 48 pounds, and the corn ensilage at $1.25 per ton. The prices for the other foods fed were the same as in the previous year.

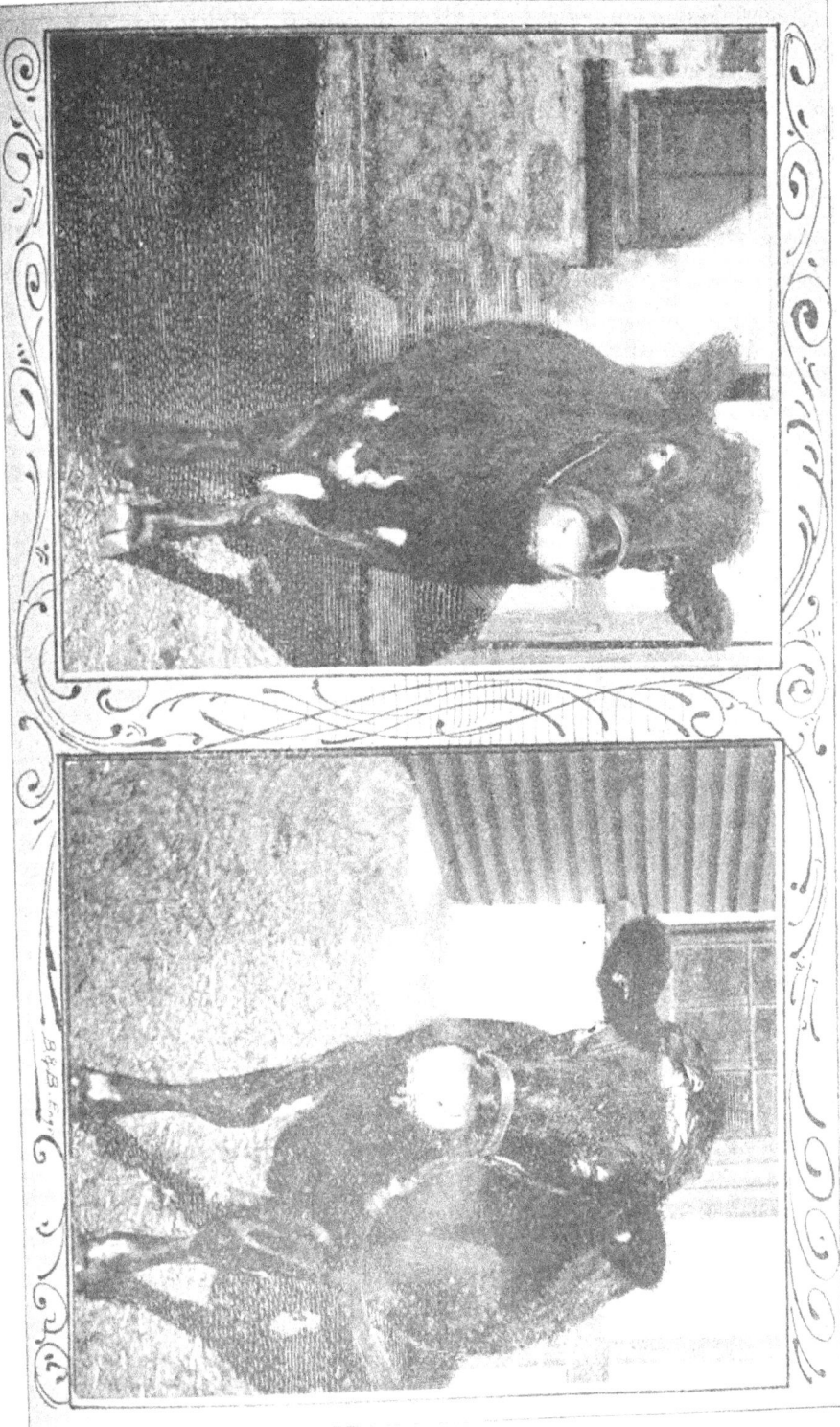

Fig. 3. Front View of the Steers Jack and Prince.

In the last six months of 1897 and the first six months of 1898, prices were materially advanced. Bran was charged at $7.50 per ton, oats at 17 cents per bushel, barley at 20 cents, corn at 22 cents, oilcake at $22.00 per ton, hay at $4.00, corn ensilage at $1.50, green food at 80 cents and mangels at 5 cents per bushel. The pasture was charged at $1.00 per month, except while grain was being fed, when it was charged at 75 cents. The green food consisted mainly of corn and sorghum.

The prices charged for some of these foods were less than the amounts paid for them. For instance, the hay purchased cost probably $2.00 per ton more than was charged for it. On the other hand, wheat bran bought at $4.50 per ton, was charged at $6.50. It was the aim to charge average market values throughout the State, as nearly as these could be ascertained, as already mentioned.

Character of the Growth Sought.—Up to the commencement of the finishing period it was the aim to keep the steers growing nicely without undue fatness. This was accomplished by feeding nitrogenous foods, and the preference was given to such of them as were cheap. The only period of even partial stagnation that occurred, was when the steers were on pasture. This unlooked for result is again referred to in that section of the bulletin which treats of pasture. During the finishing period which covered say the last 3½ months of the experiment, the process was reversed. The aim then was to fatten quickly by feeding carbonaceous foods, and as much as the steers could take with a relish. Quick and profitable fattening was made possible by the condition of good thrift in which the steers had been kept, a result that is practically unattainable when animals are unthrifty or lean when feeding for the block begins. Men are slow to learn that in growing meat, periods of stagnation in development are indeed expensive, though not equally so at all seasons of the year. Constant growth

FOOD CONSUMED.

upon suitable foods, has also a beneficient influence on the quality of the meat, as exemplied in this experiment.

Food Consumed.—Table I. gives the amounts of milk and of meal consumed by the steers respectively, until one year old.

TABLE I.

	New Milk.	Skim Milk.	Oilmeal.	Bran and Grain.
Jack	Lbs. 172	Lbs. 3,383	Lbs. 56¾	Lbs. 1,088
Prince	248	3,181	49¼	998

They were fed new milk for 12 and 15 days respectively, and on an average of 12 pounds per day. One week or a little more than that was occupied in making the change from all new milk to all skim milk. Subsequent to this period skim milk was fed to Jack for 185 days, and at the average rate of 20 pounds per day. Skim milk was given to Prince for 166 days, and at the average rate of 18.7 pounds per day. The skim milk was thus fed for a period unnecessarily long, but it was given for the reason that it was plentiful at the station. Calves may be nicely reared on a quantity of skim milk considerably less than that fed per day in the experiment, and the milk may be witheld from them when necessary, ere they reach more than half the age of the calves under consideration, that is to say, it may be witheld at an age not much beyond three months. But when the milk is at all limited in quantity, care must be taken to give free access to ample supplies of good pure water.

The oilcake was fed in the form of meal. It was not given at all until the calves were about two months old, and yet they made good gains. It was given along with other meal, only a small quantity was fed at the first, not more than a teaspoonful a day, and during no part of the year was more than ¼ of a pound fed per day. The oilmeal was not fed while the animals were getting green food.

Ground flax would probably have given even better results during the continuance of the milk period, but oilcake was chosen because it can be more easily obtained by the farmers.

The feeding of meal, other than oilcake began simultaneously with the change from new milk to skim milk, and was continued through the remainder of the year. The meal consisted of bran and oats, and it was fed practically in the proportions of say 1 and 3 parts respectively During the first part of the milk period they were virtually given all the meal they would eat with a relish, but none was allowed to remain uneaten. Subsequent to that period they would have eaten more, but in feeding the meal, the precaution was observed to give only enough to produce free and continuous growth, without inducing undue fatness. The average amount fed to Jack per day after the meal feeding began, was 3.14 pounds, and to Prince 2.78 pounds. The quantity fed never exceeded 4 pounds per day.

Table II. gives the quantities of food other than milk and meal consumed by the steers until one year old.

TABLE II.

	Hay.	Green Feed.	Corn Ensilage.	Mangels.
	Lbs.	Lbs.	Lbs.	Lbs.
Jack	951	2452	232	1651
Prince	1105½	3418	337	780

It will be observed that Prince consumed more hay and other fodder than Jack, but that the latter consumed more mangels, and by consulting the above table, it will be seen that Jack consumed more meal than Prince. These differences arose in part at least from the individual preferences of the steers, to which a due regard was had throughout the whole duration of the experiment. But it was also caused to some extent, as in the case of feeding the mangels, by the season of the year at which the food was accessible. The feeding of the mangels commenced almost simultaneously with the feeding of the grain, and continued while this food was in season. The mangels were fed sliced and in two feeds per day, and in no instance was more than 8 pounds fed per day.

The steers were not put out on pasture the first summer, and partly for the reason that none was conveniently accessible. But no special anxiety was felt with reference to such pasture, as it is yet an unsettled question as to whether calves for beef can be raised better and more cheaply on pasture, than on green food and adjuncts fed to them, where they can be kept reasonably cool and comfortable all the while.

Table III. gives the respective amounts of meal consumed by each steer between the ages of one and two years.

TABLE III.

	Oilcake.	Bran.	Oats.	Barley.	Corn.
Jack	Lbs. 33	Lbs. 548	Lbs. 825	Lbs. 20	Lbs. 187
Prince	......	482¾	658	......	184

It will be observed that the proportion of the oats to the bran fed was much less the second year than the first. It was thought that oats were less necessary than when the animals were younger, and bran was also the cheaper food. The average amount of the mixture fed to Jack per day during that part of the year that meal was being fed, was 5.19 pounds, and to Prince, 5.27 pounds, and at no time was either animal fed more than 7 pounds per day. The first was out on pasture for two months before he reached the age of two years, and the second was out on the same for four months. More or less meal was fed during the whole of the second year, except when the steers were out on pasture, as stated more minutely in a subsequent paragraph. No oilcake or barley was fed to Prince during the second year, and only a small quantity of either was given to Jack. This arose from the fact that he was nearly two months older than Prince. Though both were receiving the same kinds of food, those charged against the two closing months of each year in the case of Prince, would be charged against the two first months in the following year in the case of Jack. This explanation also applies to the other apparent variations in food given, as, for instance, green food and pasture.

Table IV. gives the respective quantities of food other than meal consumed by the steers between the ages of one and two years.

TABLE IV.

	Hay.	Green Feed.	Corn Ensilage.	Mangels.
	Lbs.	Lbs.	Lbs.	Lbs.
Jack	2,401	1,454	1,313	3,136
Prince	1,966	....	1,219	3,291

It will be observed that during the second year Jack consumed considerably more of all kinds of food than Prince, except roots. Both were fed mangels from September 28th, 1896, to May 11th, 1897. The average amount given to Jack per day was 13.94 pounds, and to Prince, 14.62 pounds. At no time did the quantity fed to either exceed 20 pounds per day. The former consumed over 22 per cent. more of hay than the latter.

Table V. gives the respective amounts of meal consumed by the steers after reaching the age of two years.

TABLE V.

	Oilcake.	Bran.	Oats.	Barley.	Corn.
	Lbs.	Lbs.	Lbs.	Lbs.	Lbs.
Jack	259½	399½	75½	404½	426½
Prince	229	342½	49	357	381

The finishing period began October 11th and ended January 31st, 1898, the last date at which the steers were weighed previous to being slaughtered. On September 7th when they were taken in from the pasture, each steer was being given 4 pounds of meal per day, and this amount was

FOOD FED.

gradually increased. On October 11th each was given 6 pounds of meal a day, consisting of bran, oats, corn and oilcake, fed in the proportions of 5, 2, 2 and 1 parts. On October 25th, the meal was changed to bran, barley, corn and oilcake, fed in the proportions of 3, 3, 3 and 1 parts respectively. And on November 15th the bran fed was reduced to 2 pounds, and the oilcake increased to 2 pounds instead of one. The quantity fed was increased gradually until December 28th when Jack was given 15 pounds of meal daily, and this amount was continued until the steers were sent to the block. On December 28th the grain portion given to Prince had been led up to 13 pounds per day. It had then to be reduced somewhat for a few days. It was again raised to 13 pounds, but could not always be maintained at that amount, and in no instance did it exceed 13 pounds. At no time of the finishing period could either steer be made to consume more than 15 pounds of meal per day, notwithstanding that a certain percentage of the same was bran, and that corn did not at any time form more than 33 per cent. of the whole meal portion fed.

Table VI. gives the amount of food other than meal that was consumed by the steers after they had reached the age of two years.

TABLE VI.

	Hay.	Corn Ensilage.	Mangels.
	Lbs.	Lbs.	Lbs.
Jack	1063	1086	698
Prince	1003	1082	694

The corn ensilage, fed from November 1st onward, the date of opening the silo, averaged 12 pounds per day, and the mangels fed averaged 5.3 pounds per day for each animal. These were given more to keep the system in tone than for any other purpose. During the finishing period the meal was the main reliance for fattening in the absence of good hay.

Details of Feeding.—The hay was fed three times a day and only what they would eat clean. But if poor hay was fed they were not forced to eat it closely during the fattening period. The residue was removed and fed to other animals. The meal was fed in two feeds per day. While ensilage was being fed, the meal was given along with the ensilage, at other times it was fed directly. The mangels were sliced in the root cutter, and were fed twice a day except during the fattening period, when they were fed but once. The steers were kept loose in box stalls during the first year, but after that time they were tied when in the stables. Water was given to them twice a day. During the season of housing they were turned into a yard each fine day from a few minutes to several hours, and salt was kept in a trough in the yard where they could help themselves.

Cost of Food.—Table VII. gives the cost of the food fed during each period, and the increase in weight made by each animal.

TABLE VII.

	Cost of Food the First Year.	Increase in Weight.	Increase in Weight per Day.
Jack	$15 23	Lbs. 670	Lbs. 1.84
Prince	14.81	693	1.90
	Cost of Food the Second Year.	Increase in Weight.	Increase in Weight per Day.
Jack	$16.43	Lbs. 400	Lbs. 1.09
Prince	14.84	402	1.1
	Cost of Food the Third Period.	Increase in Weight.	Increase in Weight per Day.
Jack	$13.51	Lbs. 365	Lbs. 2.03
Prince	10 56	228	1.81

It will be observed that the two steers made an average daily increase in weight of 1.87 pounds the first year. This increase includes the weight at birth. During the second year the average increase was only 1.09 pounds. This was certainly a poor showing. But in all experimental work common fairness demands that the bitter shall be given along with the sweet. It was caused by the miserable gains made while the steers were on pasture, to which further reference is made below. During the third period, Jack was fed for 180 days, and Prince for but 122 days.

Behavior While on Pasture.—The steers were out on pasture 127 days during 1897, that is to say, from May 24th until September 27th. When first turned on the pasture the aim was to give them grain for 14 days, by reducing the quantity gradually from 7 pounds until no grain would be given. But Prince would not eat any grain worth mentioning after he was turned out on grass. No grain was given again until September 6th, or 105 days from the date of turning onto the pasture. Grain was again given to both steers during the 21 closing days of the pasturing, but only in small quantities.

At the end of 105 days from the date of turning out on pasture, Jack had made an advance in weight of only 69 pounds, or .63 pounds per day. A small amount of grain was then given during the remainder of the pasturing period as previously stated, but which at no time exceeded 4 pounds daily and yet the daily increase in weight was more than 1½ pounds. The total increase in weight for the 127 days on pasture was 101 pounds, or .82 pounds per day. With Prince the results were still more unsatisfactory. During the first seven days on pasture, he decreased in weight from 1,018 pounds to 955 pounds, a loss of 63 pounds. During the next 7 days this loss was further increased to 73 pounds. At the end of 84 days from the date of turning out he was still 3 pounds lighter than when first put upon the pasture. After that time he began to gain slowly until grain was again given on September 6th. He then gained 20 pounds in the remaining three weeks while on pasture, or .95 pounds per day. The entire increase while out on pasture was 97 pounds, or .63 pounds per day.

These results are not a little extraordinary. The pasture was plentiful and succulent. It was blue grass, which is considered one of the best for fattening. Such pasture is looked upon as being a perfect food. A gain of nearly 2 pounds per day was looked for, while the actual average daily gain was but little more than ¾ of a pound per day. The reasons are not quite clearly apparent. The true explanation, it may be, will be found in one or both of the following reasons: First, the grass was more than usually succulent, because of the excessive rainfall. The gains made by the sheep under experiment in an adjoining field were much less during the summer of 1897 than during that of 1896 or 1895, although the pasture was more abundant in 1897 than in either of the preceeding years. And second, it may be that when animals are fed a small grain portion daily and continuously for many months, a habit of digestion is engendered which calls for the continuance of such food.

Financial Summary.—The steers were sold to the supply department of the School of Agriculture. They were thus disposed of to enable Mr. Andrew Boss, the farm foreman, to secure data and illustrations regarding the relative outcome from the different cuts of the respective carcasses. The information thus obtained will be used by Mr. Boss in illustrative work in the butchering department. The steers were valued at the prices they would sell for in the live stock markets of St. Paul and Minneapolis. The values were fixed by two experts from South St. Paul. One of these, Mr. C. Engemoen, is live stock salesman for E. M. Prouty & Co., and the other, Mr. J. C. Crosby, is the live stock buyer for Swift & Co., at South St Paul. The price put upon Jack was $4.75 per 100 pounds, live weight, and upon

Prince $4.40. Both valuators were of the opinion that the difference in price was placed at the extreme limit, and this opinion was well sustained by the behavior of the steers on the block as is shown below.

Including the value at birth, the entire cost of growing Jack was $45.17, and of growing Prince $40.21. The shrunk weights, with a shrinkage of 3 per cent. were 1,392 pounds and 1,280 pounds respectively. The first was sold for $66.12, and the second for $56.32. The profit on the first therefore was $20.95 and on the second $16.11. The items of interest on the money invested for food and in the plant are not taken into the account, nor was the cost of grinding the grain included. It was ground on the farm. The manure is supposed to offset the cost of labor and of bedding, and it should be borne in mind that the food was charged at average market values. The difference between these values and the cost of producing the food by the farmer should more than offset the interest on the money invested for food and plant while the steers were being reared.

Behavior on the Block.—In comparison with the shrunk live weight, the dead weight of Jack's carcass was 57.7 per cent. and that of Prince's 58.2 per cent. This was not in keeping with the general opinion expressed with reference to the steers before they were slaughtered, nor was it in keeping with what one would naturally expect from the conformation of the steers. But facts are facts, and cannot be gainsaid. To distort them with a view to make them fit preconceived theories in experimental work, is nothing short of a crime. The meat of both carcasses was simply superb in quality. Throughout the loin and rib cuts and also in other parts of the carcass, the admixture of the fat and lean, as brought out in the engraving, was simply perfect. Particles of fat of more or less size flecked the loin in all the best cuts, and in a way that the writer has never excelled. And this opinion was sustained by the juicy, tender

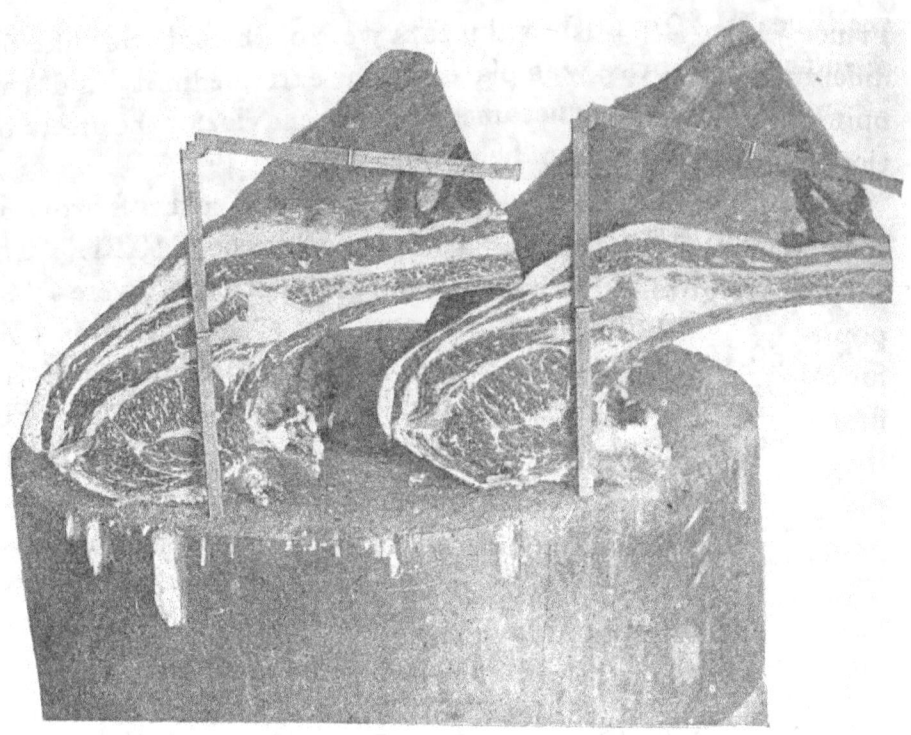

Jack.   Prince.
Fig. 4. Best Rib Roast.

and delicious character of the meat when served on the table. But little fat was laid on externally, and the per centage of internal fat was not high. The outcome in the carcasses was not far different. But all things considered, in Jack's carcass the covering of flesh on the bone was deeper than in the case of the other.

Observations—

1. It will be observed that no oilmeal was given to the calves until they were about two months old, that is to say, for about a month after the new milk ceased to be fed to them, and yet the daily gains made during the second month were 2.03 pounds on the average. This so far indicates that calves can be made to grow nicely on skim milk, bran and oats and good hay, without the aid of oilmeal at all, though they are likely to grow better when it is fed to them.

OBSERVATIONS.

2. The greater capacity of Jack to stand forced feeding during the finishing period is worthy of notice. This capacity would seem to arise from greater inherent vigor of digestion, and was indicated in the outward form, more especially in the more compact build, the wider breast, and the better spring of rib. There was much more difficulty in getting Prince to take 13 pounds of meal per day than to get Jack to take 15 pounds, this was decidedly to the advantage of the latter, as shown in the greater gains which he made per day and also in the more complete ripeness of the carcass.

3. It is also worthy of notice that during the finishing period the steers, as already stated, could not consume more than 13 and 15 pounds respectively of meal per day. To this may be added some grain in the ensilage, but not a very large amount, probably not more than a pound a day. This would seem to confirm the view, that when cattle are fed from 25 to 30 pounds of grain per day during the finishing period, much of the grain so fed is wasted, even though pigs are made to glean amid the droppings.

4. The difference in the profit made by a small variation in the price paid, is well brought out in this experiment. Valuing the carcase of Jack at $4.75 as was done, and that of Prince at $4.40, the profit on the first exceeds that on the second by $4.84. Had Prince also been valued at $4.75 per 100 pounds, that excess of profit would have been reduced to 40 cents.

5. But for the low gains made by the steers when on pasture, the financial outcome would have been very much better. The entire gain of the two steers while out on pasture for 127 days was 198 pounds; a normal gain at such a time would have added fully 250 pounds more to the weight, which would have materially enhanced the profit from growing the steers. Hence the great loss from periods of stagnation in development when meat is being grown on the intensive plan when these occur.

6. Since the food fed was charged at the average market values in the state, the profit would have been considerably increased had it been charged at the cost of production. Such cost would be the correct basis of values could it be ascertained.

CONCLUSIONS.

1. Since the steers were sold for $37.06 above the cost of production, under the conditions stated, it is just to conclude that a good profit can be made from growing beef in Minnesota, even when grown on the intensive plan.

2. That the method of growing meat thus is applicable to average farm conditions in the state, since the foods fed may all be produced on the farm, except the bran and oilcake.

3. That the quality of meat thus grown is of the very best, and should therefore command the highest price in the market.

4. That in growing beef as in this experiment, the farmer can get much better values for the food products which he grows by feeding them at home, rather than by selling them directly.

FATTENING RANGE STEERS IN WINTER.

SECTION NO. 2.

THOMAS SHAW.

It was intended that this experiment should be conducted on substantially the same lines as that of the previous winter which also experimented with feeding range steers, as detailed in Section 2 of Bulletin No. 58. But as will be shown in the proper place, it was found necessary to modify the whole plan of procedure, and for reasons that will be given. This experiment therefore is not a direct confirmation of that of 1896-97 as it was intended to be, in the sense of determining the relative results from feeding smaller and larger quantities of meal. It is confirmatory of the former however in so far as it throws light on the financial aspect of the question, and, notwithstanding the change of plan that was found necessary, the experiment presents some interesting and valuable features. But these will not be anticipated. In the proper place they will all be stated.

Time Covered by the Experiment.—The steers were put on full feed November 15th, 1897. They were sent to the block May 28th, 1898. The period of feeding therefore lasted for 194 days. On November 6th they were tied in the stall, and during the intervening days until the experiment proper began, they were gradually brought up to a full ration.

The Objects of the Experiment—As at first planned the chief of the objects sought in the experiment were the following: 1, to ascertain the comparative results from fattening steers on smaller and larger quantities of meal; 2, to gather information as to the behavior of range steers while under full feed, and 3, to learn the financial results. Chief among the secondary objects were the following: 1, to learn the daily gains that would be made from a certain line of feeding; 2, to learn as to the consumption of food fed in producing these gains, and 3, to learn the cost of the same. One of the steers however had to be taken out of the experiment before it was fairly under way. He manifested a disposition so vicious and sullen that to continue him in the experiment did not promise any adequate results. He was exchanged therefore for another. A second steer manifested a shy and sullen disposition. When tied up he threw himself violently and for a good while loudly proclaimed his fancied wrongs by bellowing. As there was no other suitable steer on hand to supplant him, he was retained in the experiment until the end thereof. He continued shy and played the Ishmaelite in his disposition throughout the feeding period, and as a result made but poor gains. About the middle of the feeding period therefore the feeding of light, heavy and intermediate quantities of meal, as described below, was discontinued. The steers were all put upon what may be termed a pretty stiff grain portion, to ascertain if the daily gains would not be stimulated by such feeding. But as the results will show, the increased gains did not materialize. And just here it may be stated that the rebellious steers would probably have fattened fairly well in the open feed lot, though not so well as those of a better disposition. This remark is interjected here, lest the conclusion should be reached that range steers are not suitable material for being fattened. The behavior of these steers does not prove that. It simply proves that they are not the best of material for being fattened in the stall. While in our judgment they ought to be fattened loose and in a shed with access to a yard, such a method of fattening would not of course have given us the results that we sought.

The Animals Used.—Nine steers were fed in all. They formed part of a carload of cattle purchased by the Hon.

ANIMALS USED. 23

W. M. Liggett, the director, and the writer, on the range. The cattle were bought from John Manning, of Shelter Ranch, some 12 miles northward from Culbertson, in Montana, on the line of the Great Northern railroad. Seventeen animals were purchased. Some were young females that had failed to breed and some were native steers of fine finish. The cows and native steers were bought to be used, first, in demonstrative work in the slaughter house, and, second, on the dining tables of the School of Agriculture. They made a fine quality of meat. Four more animals owned by Mr. Manning completed the carload shipped to the Minnesota Transfer.

They reached the experiment farm October 20th, and were kept on blue grass pasture supplemented by some grain until November 6th, when the nine animals put in the experiment were weighed separately, and were tied in stalls side by side. The freight on the car from Culbertson to the Minnesota Transfer, 702 miles, including $2.20 for food, was $87.80. The proportion of the freight paid on the nine steers on the basis of weights was $28.80, or $3.20 on one steer.

In age the nine steers were two years old, but evidently there was no little variation in the age, as there was considerable variation in the size. In breeding, the shapes, the color markings, the presence or absence of horns, and the character of the horns when present, indicated Shorthorn, Hereford and Aberdeen Poll blood. They had been sired by high grade or pure sires of one or the other of these breeds, and in individuality they were good, but a little more leggy than the really typical beef steer ought to be. Unevenness in size and dissimilarity in disposition were the chief points of weakness in the selection.

In determining the cost of the steers when the experiment proper began, November 15th, they were valued at the average price per pound paid for the whole lot when purchased on the range. No other course was open upon which to base valuations, as they were not priced individually at the time of purchase. The proportionate cost of freight and food were added to this sum. They were also given the advantage accruing from the handling of the other animals in the carload lot. Taking these figures as the basis of the total outlay, the total cost of the nine steers up to November 15th was $261.58, and the cost per 100 pounds, on the basis of weight when they entered the experiment, was $2.69½ cents, a price considerably lower than the average market value of such animals at the time.

Conditions Governing the Experiment.—The steers were divided into three lots, with three in each. Those in lot 1 were to be fed the light meal portion, the steers in lot 3 the heavy meal portion, and those in lot 2 the intermediate quantity. As it was thought that weighing the steers so frequently the previous winter tended to disturb them unduly, it was determined that they should be weighed but once while the feeding was going on, in addition to the inevitable weighing at the beginning and at the close of the experiment. This intermediate weighing was made on March 7th, 112 days after the experiment began. They were of course not turned out to exercise. They were curried occasionally, at the first, with a brush on a handle and at a safe distance. Some of them, however, became quite tractable and responded nicely to kind treatment, while others continued to resent close handling until the end. And it is significant that the steers which responded best to handling made the best gains.

Food and Feeding.—The meal fed to the steers for the first 112 days of the experiment was composed of bran, barley and corn. The first period of 28 days it was fed in the proportions of bran 5 parts, barley 3 and corn 2. The second period of similar duration it was changed to bran 4 parts, barley 3 and corn 3. The third period it was made up of 3 parts bran, 3 barley and 4 corn. And in the fourth period it was further changed to 2 parts bran, 3 parts barley and

5 parts corn. These proportions are by weight. The aim in thus changing the food frequently was to increase gradually the more carbonaceous and more concentrated factor of the food, viz. the corn. At the commencement of the experiment, the steers in lot 1 were fed 6 pounds per day of the mixture, the steers in lot 2, 8 pounds, and the steers in lot 3, 10 pounds. These respective amounts were increased 1 pound at the end of every four weeks, and during the first 112 days of the experiment 1 pound of oilcake in the nutted form was given to each steer per day. When the steers were weighed on March 7th, it was apparent from the behavior of the steer No. 4 that the experiment was so marred, that in one of its essential objects, viz. comparison of the results from the light and heavy grain feeding, it would be only prudent to abandon it. It was determined at the same time to try and ascertain if heavy feeding during the remainder of the term would stimulate greater gains. The meal portion was changed to bran and corn in the proportions of 1 and 3 parts, and it was decided to feed all the steers up to their capacity for consumption without putting them off their feed. The 1 pound of oilcake per animal was also continued and during the last 25 days of the experiment it was changed to 2 pounds.

Mixed hay, timothy and clover, of medium quality, was fed in quantities such as the steers would consume. And after 112 days of feeding, corn ensilage was added. Of this however they only consumed from 6 to 8 pounds per day, and one steer, viz. No. 6, would not take any. The grain was ground and fed directly, in two feeds per day. The hay was given in the uncut form and in three feeds per day. And the ensilege was fed in two feeds.

Estimated Value of the Food.—The food was estimated at what may be termed approximate average market values for the state. They were as follows:

Bran, per ton.. $7.50
Barley, per bushel of 48 pounds.................................... .18

Corn, per bushel of 56 pounds... .22
Oilcake, per ton.. 22.00
Hay, mixed, per ton... 4.00
Corn ensilage... 1.25

Five cents per sack is the common charge for grinding coarse grain. Adding the cost of grinding to the barley and corn, makes the cost of these 20½ cents and 24½ cents respectively per bushel.

These prices were considerably in advance of those allowed for nearly all the products during the three previous years. The most notable advance, however, is oilcake, which went up in a single season from $14.00 to $22.00 per ton. The advance in these values of course add to the cost of the feeding and in the absence of higher prices for the finished product, correspondingly reduces the profit that would otherwise have accrued.

Food Consumed.—Table VIII. gives the total amount of hay, ensilage and meal consumed by each steer during the experiment and the sum of these taken together.

TABLE VIII.—Food Consumed by the Steers.

	Hay.	Ensilage.	Meal.	Total.
	Lbs.	Lbs.	Lbs.	Lbs.
No. 1	1,965	633	2,132	4,730
No. 2	1,552	637	2,356	4,545
No. 3	1,847	639	2,615	5,101
No. 4	1,380	640	2,100	4,120
No. 5	1,656	598	2,331	4,585
No. 6	1,621	......	2,665	4,286
No. 7	1,871	594	2,123	4,588
No. 8	1,912	641	2,371	4,924
No. 9	1,609	635	2,588	4,832
Total	15,413	5,017	21,281	41,711

The variable quantities of food consumed are not to be attributed entirely to variations in the capacity of the steers to consume, as, during the first 112 days of the experiment, they were fed varying quantities of meal.

WEIGHTS AND INCREASE.

Table IX. gives the quantity of each food factor consumed daily by each individual steer throughout the experiment, and also the total daily consumption of food by each.

TABLE IX.—Daily Consumption of Food by the Steers.

	Hay.	Ensilage.	Meal.	Total.
	Lbs.	Lbs.	Lbs.	Lbs.
No. 1	10.13	3.26	10.99	24.38
No. 2	8.00	3.28	12 14	23.42
No. 3	9.52	3.29	13.48	26.29
No. 4	7.11	3.30	10.82	21.23
No. 5	8.54	3.08	12.02	23.64
No. 6	8.35		13.74	22.09
No. 7	9.64	3.06	10.94	23.64
No. 8	9.86	3.30	12.22	25 38
No. 9	8.29	3.27	13.34	24.90
Average	8.83	2.87	12.19	23.89

In the above table we are given an approximate idea of the daily consumption of food by steers weighing as these did in the experiment, when fed practically on a dry ration. The average weight of the nine steers during this period of feeding was 1178 pounds. The consumption of meal was a little more than 12 pounds per day. It is certainly under 13 pounds allowing for what might be in the ensilage. This amount is far short of what is fed on an average in the feed lots of the west. During the first 112 days of the experiment some of the steers would have consumed more grain if it had been given to them, but it is more than questionable if increased gains would have resulted, as will be shown later. The difference in the amounts of food consumed per day is relatively much less than the difference in the proportionate gains resulting from them.

Weights of the Animals.—Table X. gives the weights of the individual animals at the commencement of the experiment and at its close and also the total individual increase and the average daily individual increase made during the same.

TABLE X.—Weights and Increase.

	Weight on Nov. 15, 1897.	Weight on May 28, 1898.	Total Individual Increase.	Daily Increase.
	Lbs.	Lbs.	Lbs.	Lbs.
No. 1	1100	1445	345	1.78
No. 2	840	1155	315	1.62
No. 3	1075	1520	445	2.29
No. 4	950	1105	155	.80
No. 5	1065	1275	210	1.08
No. 6	1075	1315	240	1.24
No. 7	960	1220	260	1.34
No. 8	1125	1440	315	1.62
No. 9	1150	1380	230	1.19
Average	1038	1317	279	1.44

There are some things in the above table that are calculated to somewhat perplex any one who has made a study of the feeding question. First, there is the uncommon disparity in the gains made by individual steers. Leaving out No. 4 for the present, it will be noticed that three steers, Nos. 1, 3 and 8, made together a daily increase of 5.69 pounds, while the three steers Nos. 5, 6 and 9 made together a daily increase of but 3.51 pounds. The first three consumed together 36.69 pounds of meal daily, and the last three 39.10 pounds. The first three weighed together 3,300 pounds when the experiment began, and the last three weighed together 3,290 pounds at the same date.

Second, the difference in type in the steers does not sufficiently account for the disparity in the individual gains. Of the six steers mentioned above, No. 9 only was pronouncedly off in type. No. 3 made an increase of 2.29 pounds per day, an uncommon increase for so long a period of feeding. No. 9 which consumed practically the same amount of meal gained only 1.44 pounds per day. This however was looked for from the pronounced difference in the type of the steers. No. 5, taking 12.02 pounds of meal per day on an average, made a daily gain of but 1.08 pounds as compared with 2.29 pounds, less than half as much as was produced by the steer No. 3, which took but 13.48 pounds of meal per day. Here the types were not far different. Some judges preferred the form of No. 5 when the feeding began. He took his feed regularly, but who can tell what he did with it? It was apparently digested so far as one could judge by the droppings. No. 4, the nervous and ill-natured steer only gained .80 pounds per day. The reason in this case is clear. He would not eat enough, nor would he rest. But why did No. 6 gain only 1.24 pounds per day, though he consumed more meal than No. 3. The type in the case of No. 6 was not much inferior if any to that of No. 3.

From such behavior we are driven to the conclusion as one result of this experiment, that we must not rely too much upon type. While we can make a grand good use of it in selecting animals for feeding as in selecting them for the dairy, we must not stake everything on it. In both there are some exceptions. The inheritance of certain other things would seem to be quite as important as the inheritance of form. In some instances at least some subtle influences lie away down in the digestive system of an animal which baffle preconceived judgments as to the outcome. Did we know more about the inheritance of the animals we would doubtless be less at sea. In the meantime we do wisely when we remember that in selecting animals, form does not tell all the story of their fitness for a certain end.

Third, the absolute average gains made were low, notwithstanding that they were excellent in at least three instances. With the steer No. 3 they were superior to any gains that have yet been realized at this station for so long a period of feeding. Had the gains made by all the steers been low, we would have sought for the result in some untoward conditions regarding food or management, or both, but the marked well-doing of some of the steers sets this at rest. That the steers made an average gain of but 1.44 pounds per day must, therefore, be owing to the poor feeding quality in some of the steers.

And just here it may be mentioned that increasing materially the proportions of concentrated foods fed did not mend matters. During the first 112 days of the experiment, the steers made an average daily gain of 1.50 pounds. To produce this gain they were fed daily on an average, 10.26 pounds of hay and 10.49 pounds of meal. Subsequently they made a daily gain of but 1.36 pounds, though given each day 6 06 pounds of hay, 6.79 pounds ensilage, and 14.51 pounds of meal. In fact, during those 82 days they were given all the meal that they could stand. Of course, during the last part of a term of feeding, animals require more food to make equal gains, but had it been a fact that some of the steers were being fed too little meal before March 7th they should have gained more rapidly subsequently, which they did not. To illustrate, take the three steers that were fed a light meal portion up to March 7th, viz., Nos. 1, 4 and 7. Previous to that date they were fed daily on an average 8.48 pounds of meal, and gained 1.31 pounds each per day. Subsequently they were fed on an average 14.25 pounds of meal each per day, and made an average daily gain of 1.30 pounds.

Table XI. gives the cost of each food factor fed to the individual animals, and also the cost of these taken together.

TABLE XI.—Cost of Food Consumed.

	Hay.	Ensilage.	Grain.	Total for Each Steer.
No. 1	$3.93	$0.79	$10.42	$15.14
No. 2	3.10	.80	11.34	15.24
No. 3	3.69	.80	12.43	16.92
No. 4	2.75	.80	10.28	13.83
No. 5	3.31	.75	10.23	14.29
No. 6	3.24		12.65	15.89
No. 7	3.74	.74	10.38	14.86
No. 8	3.88	.80	11.41	16.09
No. 9	3.22	.79	12.31	16.32
Total	$30.86	$6.27	$101.45	$138.58

A noticeable feature of this and indeed of all feeding for the block is the marked contrast between the cost of the fodder and that of the grain. The former cost $37.13 and the latter 101.45, or nearly three times as much, hence the wisdom of trying to utilize fodder as far as may be consistent when fattening animals.

Table XII. gives the average daily cost of the food consumed by the individual animals during the first 112 days of the experiment, during the last 82 days of the same, and during the whole experiment.

TABLE XII.—Average Daily Cost of Food Consumed.

	For the First 112 Days.	For the 82 Last Days	For the Whole Experiment.
	Cts.	Cts.	Cts.
No. 1	6.50	9.46	7.80
No. 2	6.96	9.07	7 86
No. 3	8.09	9.59	8.72
No. 4	5.79	8.96	7.13
No. 5	6.17	8.99	7.87
No. 6	7.58	9.02	8.19
No. 7	6.54	9.18	7.66
No. 8	7 37	9.49	8.27
No. 9	7.77	9.29	8.42
Average	6.97	9.23	7.94

The average daily cost of the food consumed rises from 6.97 cents per animal per day, during the first period of 112 days of moderate feeding, to 9.23 cents per day during the second period of 82 days of forced feeding. That the daily cost usually rises with the advancement of the fattening period, and that the daily gains usually decrease somewhat, is common experience. But in this experiment the relative rise is more than it would have been under a more moderate system of feeding. The ration fed was, therefore, unnecessarily costly, while the forced feeding was in process, and the gains resulting were probably less than they would have been under more moderate feeding.

Cost of Increase.—Table XIII. gives the cost of making 100 pounds of increase by each animal during the first 112 days of the experiment, during the last 82 days of the same, and during the entire experiment.

TABLE XIII.—Cost of Increase per 100 Pounds.

	For the first 112 days.	For the last 82 days.	For the whole period, 194 days.
No. 1	$4.03	$4.79	$4.39
No. 2	3.90	6.47	4.84
No. 3	3.29	4.62	3.80
No. 4	8.31	9.54	8.92
No. 5	5.20	9.57	6.80
No. 6	5.74	8.04	6.62
No. 7	4.12	9.18	5.71
No. 8	4.13	6.77	5.11
No. 9	7.25	6.93	7.10
Average	$5.11	$7.32	$5.92

The wonderful difference of individuality in animals that are being fattened, in turning the food fed to them to account, is well brought out in the above table. While the steer No. 3 made 100 pounds of increase at a cost of $3.80, the steer No. 4 made the same at a cost of $8.92.

Profit Made.—Table XIV. gives, 1, the value of each individual steer when the experiment began; 2, the cost of the food fed; 3, the total outlay; 4, the value of each animal when the experiment closed; and 5, the profit made on each. The totals are also given in each instance.

TABLE XIV.—Values and Profit Made During the Experiment.

	Value on Nov 15, when the experiment began.	Cost of Food.	Total Cost.	Value on May 28, '98, when the experiment closed.	Profit.
No. 1	$29.65	$15.14	$44.79	$66.83	$22.04
No. 2	22.64	15.24	37.88	53.42	15.54
No. 3	28.97	16.92	45.89	70.30	24.41
No. 4	25.60	13.83	39.43	51.11	11.68
No. 5	28.70	14.29	42.99	58.97	15.98
No. 6	28.97	15.89	44.86	60.82	15.96
No. 7	25.87	14.86	40.73	56.43	15.70
No. 8	30.32	16.09	46.41	66.60	20.19
No. 9	30.99	16.32	47.31	63.83	16.52
	$251.71	$138.58	$390.29	$548.31	$158.02

The results as stated above do not take any account of shrinkage and for manifest reasons. This is done when the financial statement is given.

The profit is most satisfactory, but it does not arise from good gains, for, as previously shown, these were on the average low rather than high. The substantial profit is then the outcome of the low cost price of the steers as compared with the selling price, than of successful feeding.

While the profit made from feeding 9 steers for 150 days in 1896, that is to say, during the experiment proper, was only $43.17, as stated in Bulletin No. 58, Section No. 1, the profit from the nine steers now under consideration was $158.07, during the 194 days of the experiment proper, and yet the gains made during the former were much better than during the latter. The great influence which the price paid and received exercises on profit is thus emphasized.

The difference in the individuality of animals is well brought out in the results of this experiment. If the steer No. 3, which made the highest profit, is contrasted with the steer No. 4, which made the lowest profit, the comparison stands thus: While the former gave a profit of $24.41, the latter gave a profit of but $11.68, that is to say, not half as much. And if the three steers which made the highest gain are contrasted with the three which made the lowest gain, the comparison stands thus: While the profit on the first three aggregated $66.64, the profit on the last three aggregated but $42.92. Great is the power of individuality even in steers that are being fattened.

It may also be mentioned here that the increase made during the experiment cost $5.92 per 100 pounds. As it was sold for $4.62½ per 100 pounds, it cost more than it was worth, a result brought about, first, by the advanced cost of food, and, second, by the moderate gains resulting from feeding it. In two instances only was the increase made worth more than the food used in making it. These were with the steers in Nos. 1 and 3 respectively.

Disposal of the Steers.—The steers were sold to Peter Van Hoven, of New Brighton. They were slaughtered for home consumption, that is to say, consumption in the twin cities of St. Paul and Minneapolis. The price paid, as stated above, was $4.62½ per 100 pounds with a shrink of 3 per cent.

Financial Statement.—Cash received for 9 steers May 28th, 1898, shrunk weight 11,499 pounds, at $4.62½ per 100 pounds.. $531.83

Value of 9 steers on November 15th, 1897, on the basis of cost.................$251.71
Cost of food... 138.58

Total outlay... $390.29

Total net profit... $141.54
Net profit on one steer............................. 15.73

Observations.—1. Since the food fed was charged at the average market values in the state, these would represent more than the cost of growing the same, and would therefore be so far unfavorable to the making of profits.

2. The value of the manure is supposed to offset the cost of bedding and labor, and also the interest on the money involved.

3. Had the experiment been closed on March 7th, and had the steers been then valued, as they might justly have been, at the prices for which they were sold, viz., $4.62½ per 100 pounds, the greatest profit would have been realized from the three steers in lot 1, which were fed the light meal portion. The respective profits, excluding shrink, would have been for those of lot 1, $57.21; for those in lot 2, $54.58; and for those in lot 3, $55.59.

IMPORTANT FACTS SUMMARIZED.

Values—

1. Value per 100 pounds on the basis of cost when the experiment began, Nov. 15th, 1897...............$2.69½

IMPORTANT FACTS SUMMARIZED.

2. Value per 100 pounds on the basis of actual receipts, May 28th, 1898, when the experiment closed $4.62½
3. Advance in value per 100 pounds 1.93

Freight—

1. Cost of freight on a carload of cattle (21 animals) from Culbertson, Montana, to St. Paul, 702 miles, including food charges ($2.20)............ 87.80
2. Average cost of freight per animal on the 9 steers fed in the experiment on the basis of weights...... 3.20

Weights—

Lbs.
1. Average weight of the steers when the experiment began, Nov. 15th, 1897 1,038
2. Average weight at the close of the experiment, May 28th, 1898 1,317

Increase in Weight—

1. Average increase in weight of each steer during the 194 days of feeding 279
2. Average increase in weight per day of each steer during the first 112 days, when the feeding was moderate 1.50
3. Average increase in weight per day of each steer during the last 82 days, when the feeding was forced 1.36
4. Average increase in weight per day of each steer during the 194 days of feeding 1.44

Food Consumed—

1. Average daily consumption of meal per animal during the first 112 days, when the feeding was moderate 10.26
2. Average daily consumption of meal per animal during the last 82 days, when the feeding was forced 14.51

3. Average daily consumption of meal per animal during the entire period of feeding.................... **Lbs.** 12.19
4. Average daily consumption of food per animal during the whole period of feeding.................... 23.89

Cost of Food.—

Cts.

1. Average cost of food per animal per day during the first 112 days when the feeding was moderate 6.97
2. Average cost of food per animal per day during the last 82 days, when the feeding was forced..... 9.23
3. Average cost of food per animal per day during the whole period of feeding.......................... 7.94

Cost of Increase.—

1. Average cost of making 100 pounds of increase during the first 112 days of the experiment when the feeding was moderate........................... $5.11
2. Average cost of making 100 pounds of increase during the last 82 days when the feeding was forced.. 7.32
3. Average cost of making 100 pounds of increase during the whole period of feeding................. 5.92

Increase in Value.—

1. Average value of each steer without shrink on the basis of cost when the experiment began, November 15th, 1897............................... 27.97
2. Average value of each steer without shrink when the experiment closed, May 28th, 1898.......... 60.91
3. Average advance in value on each steer from feeding for 194 days, or, during the entire experiment... 32.94

Profits.—

1. Aggregate net profit from feeding 9 steers for 194 days.. 141.54
2. Average net profit from feeding one steer for the same period....................................... 15.73

CONCLUSIONS.

The following are prominent among the conclusions that may legitimately be drawn from the experiment:

1. That inasmuch as the steers fed on the light meal portion gave the largest profit until March 7th or as long as they were thus fed, this result is in keeping with those previously obtained from feeding steers thus at our station.

2. That because of the sullen temper and restless bearing of two of the steers when first tied up and of the ill-doing of the one of these retained in the experiment, range steers would not seem to furnish very suitable material for being finished in the stall.

3. That the difference in the capacity of animals to make increase is very great, since in this experiment, the average daily gains on the same ration in kind ranged from .80 pounds per day to 2.21 pounds per day.

4. That the great difference in the ability of animals to give profitable returns while being fattened is emphasized in the fact that the steer No. 3 made 100 pounds of increase at an average cost of $3.80, while with the steer No. 4 it cost $8.92 to secure a similar increase.

5. That while good type is one guaranty of good feeding, it alone does not furnish a sure guaranty, as witnessed in the low profits relatively from the steers Nos. 4, 5 and 6, all of which were possessed of good type.

6. That the decrease in the relative gains during the last 82 days of the experiment, with the marked increase in the cost of the same while the steers were under what may be termed a forcing ration, emphasizes the futility of attempting to force gains by excessive grain feeding.

Some Averages from Two Experiments.

The following are some of the more important of the averages obtained from feeding range steers taken from the two experiments conducted at our station in this line of work.

Values—

1. Value per 100 pounds on the basis of cost when the animals were first put under experiment........$2.91¼
2. Value per 100 pounds on the basis of receipts when the animals were sold................................. 4.62½
3. Advance in value per 100 pounds.......................... 1.71¼

Weights—

Lbs.
1. Average weight per animal when put under experiment.. 1102½
2. Average weight per animal, without shrink, at the close of the feeding period................................. 1367
3. Average increase in weight per animal per day.... 1.67½

Food Consumed.—

1. Average amount of meal consumed daily by one animal.. 11.65
2. Average amount of ensilage consumed daily by one animal.. 10.76
3. Average amount of hay consumed daily by one animal.. 9.19
4. Average amount of food consumed daily by one animal.. 31.60

Cost of Food Consumed.—

Cts.
1. Average cost of food for one steer per day........... 7.57
2. Average cost of making 100 pounds of increase... $4.94

Increase in Values.—

1. Average value when the experiment began, without shrink.. 32.24
2. Average value when the experiment closed, on the same basis.. 63.22
3. Average advance in value from feeding for an average period of 162½ days.................................. 30.98

Profit.—

1. Total net profit on 18 steers in the two experiments... 294.29
2. Average net profit on one steer............................ 16.35

FEEDING PIGS OF DIFFERENT GRADES.

SECTION NO. 3.

THOMAS SHAW.

No question pertaining to the growing of live stock is attracting so much attention to-day in the United States as that of the bacon hog. Nor is there any question in live stock circles that is provoking so much of controversy. The writer has had but one opinion on this question, viz.: That the average hog grown in the United States must be so modified in the near future as to more nearly resemble the bacon hog in form and characteristics, and that the change will have to be made, even though not one pound of bacon should be sent to the markets of Great Britain. This change would seem to be inevitable for the reason, first, that the taste of the consumer calls for leaner pork than was in demand during the recent vanishing days when the lard era prevailed; second, that it will give our pigs more bone and stamina, the latter of which is the crowning requisite of domestic animals; third, that it will greatly increase the prolificasy of our swine, and, fourth, it will accomplish all this without seriously decreasing their capacity to produce much increase relatively from a minimum of food.

There are two ways of obtaining such a result. The first is by selection, and the second is by crossing. The first is the easiest of adoption by far by the average American breeder, because of the paucity of the material for crossing, but it is considerably the slower method of the two. By the first method the swine grower must needs select sows for breeding long and deep and rangy in body, and standing on good limbs. He must needs select good, strong, vigorous boars, but less rangy than the sows. He must adapt his food to the needs of the bacon hog by making it more nitrogenous than a diet is when composed mainly of corn, and in a few generations he will have his bacon hog, and without any serious dimunition of good growing and good feeding qualities. By the second method, boars of the Improved Large Yorkshire and Tamworth breeds will be crossed upon the short-bodied sows that abound in all parts of the land, and more especially within the corn belt. The boars of these respective breeds should only be chosen for these crosses with discrimination. Those rangy beyond a certain degree should be rejected, lest excessive range in the male should impart to the progeny a lack of feeding qualities. The first cross will give the animal sought. If our people would only smash and thow away the glasses of prejudice with which they view this matter, and if they would but introduce this cross where it can be done, they would be astonished at the renovating influence it would have upon their swine in a single generation.

With the writer, experimentation with these crosses is no new thing, nor is experience in growing the Improved Yorkshire and Tamworth swine. While at the Ontario Experiment Station, these crosses were being made from year to year. Not much was published, however, regarding them, because of the severance of relations with that institution at a time when the work was yet in progress. But enough was done to convince one of the much power which both these breeds have to bring great renovation to the swine that we have, enfeebled, as many of them are, by generations of imprudent feeding, more especially in the corn belt.

Experimentation in making these crosses began at our station in 1894. This was some three years prior to the now famous utterance of Secretary Wilson on Tamworth swine. And much material would now be on hand for publication with regard to them, but for an untoward event that destroyed so much of the same as to well nigh erase the somewhat elaborate system of crosses that had been instituted. In an evil hour the pestilence that walketh in darkness swept through our herd and it left us with much of our precious treasure to bury in the form of dead swine. The dread disease hog cholera had been brought to us in the early spring of 1897, doubtless by some visitor, and we were left lamenting the loss of the work of years.

Our little experiment however escaped the wreck, and for the reason that it had been completed the previous autumn, the details of this experiment are given in the bulletin. When speaking to the farmers about these crosses, the writer had been frequently met with the objection, that while the first cross might be good, the second cross would show retrogression, and that subsequent crosses would show still further retrogression. The experiment was undertaken in the hope of throwing some light on the question raised, and more especially as to the relative merits of the progency of Improved Yorkshire sires of the first and second crosses. The direct bearing of this whole question upon the future prosperity of our State is the apology for the somewhat prolonged introduction to this bulletin, since no State in the Union possesses a higher natural adaption to the production of the bacon pig than our own, because of the great variety and abundance of the foods that may be grown.

Time Covered by the Experiment.—The experiment commenced July 13th, 1896. It ended November 2d of the same year. It thus covered 112 days. It was made up of four periods of 28 days each. It is not to be regarded simply as an experiment in fattening swine, but rather as an experiment in growing and fattening them combined.

The Objects of the Experiment.—Chief among the objects of the experiment were the following: 1, to ascertain the relative merits of swine of the first and second crosses from improved Yorkshire sires in producing growth, and for fattening; 2, to ascertain the relative values of corn and barley respectively as food for swine when fed as in the experiment; and 3, to ascertain the financial outcome under the existing conditions. And prominent among the secondary objects were the following: 1, to ascertain the relative daily increase; 2 the food used in making it; and 3, the cost of the same.

The Animals Used.—The animals used in the experiment were chosen from two litters, designated respectively as the first and second Yorkshire crosses. Six individuals of each cross were fed. Those of the first cross were from a first class pure bred improved Yorkshire boar and a high grade Berkshire sow of somewhat heavy build. In the sow there was probably a dash of Poland China blood as indicated by her large and almost pendent ears. The pigs of the second cross were by the same Yorkshire sire and out of a first cross Yorkshire dam. This dam was the progency of the sow referred to above and a pure Yorkshire sire. The grade Berkshire dam was of course black in color with white points. The first cross Yorkshire dam was pure white, a beautifully formed and handsomely developed sow.

The six first cross pigs were from the second litter produced by the dam. The litter comprised 8 living animals. They were farrowed March 27th, and were reared on the dam until fully 10 weeks old. Subsequently and until they entered the experiment, they were fed skim milk and shorts with a small proportion of oats and barley added. They had the run of a pasture of about an acre, more or less, but could get but little food therefrom, and in consequence they were given some green food, chiefly grass, which was carried to them. The six second cross pigs were the first litter produced by the dam, and they embraced the whole litter. They were farrowed March 30th, and were similar in all respects to the pigs of the first litter.

Conditions Governing the Experiment.—The pigs were divided into four lots of 3 each. Lots 1 and 3 comprised pigs of the first cross and lots 2 and 4 pigs of the second cross. Lots 1 and 2 were given the corn diet described below, with adjuncts, and lots 3 and 4 were given the barley diet with similar adjuncts. The apartments of the piggery in which they were fed were side by side, and each was 8x12 feet in area. Each lot had the run of a small paddock for an hour or more every day. The pigs were weighed at the beginning and close of the experiment, and also at the end of each intervening week.

Both litters were 108 days old when the experiment began; as the pigs of the first cross were put under experiment 3 days earlier than those of the second cross, but with them the experiment also ended 3 days sooner. The conditions of the experiment therefore were very even. The pigs were from the same sire, were farrowed practically at the same time, and were the same age when the experiment began and closed.

Food and Feeding.—The pigs in lots 1 and 2 were fed oats and corn during the first period in the proportions of 3 and 1 parts respectively by weight. During the second period the proportions of these foods were 2 and 2 parts respectively. During the third period they were 1 and 3 parts respectively, and during the fourth period corn only was fed. The pigs in lots 3 and 4 were fed similarly except that barley was substituted for corn. The grain was all ground. It was prepared by soaking in cold water for 12 hours previous to feeding, but that fed at noon was soaked for but six hours. A small amount of salt was also added daily to the food. It was fed in three feeds, and the pigs were given all that they would eat clean and with a relish. They were also given such green food as was in season. The green food comprised corn, second growth clover, rape and cabbage.

Estimated Value of the Food.—The food was estimated at the average market values in the State. These were very low at the time, but not relatively lower probably, than the price of pork.

These market values were as follows:

Oats, per bushel of 32 pounds.................................. $0.14
Barley, per bushel of 48 pounds.................................. .16

Corn, per bushel of 56 pounds.................................... $0.18
Green food, per ton... .75

An allowance of 5 cents per sack was made for grinding the grain. This is the usual charge for such work. The charges for grinding, therefore, would make the oats 16½ cents, the barley 18½ cents and the corn 20½ cents per bushel.

Food Consumed.—Table XV. gives the respective amounts of each kind of grain and also of the green food consumed by the animals of the respective lots; first, during each period of the experiment, and second, during the whole experiment.

TABLE XV.—Food Consumed by the Pigs.

	Oats.	Corn.	Green Food.	Total.
Lot 1—	Lbs.	Lbs.	Lbs.	Lbs.
First Period	17½	62½	106	346
Second Period	136	136	60	382
Third Period	100½	301½	90	492
Fourth Period		480	66	546
Total	414	980	322	1,716
Lot 2—				
First Period	197	65	130	392
Second Period	142	142	52	336
Third Period	110½	331½	30	472
Fourth Period		495	85	580
Total	449½	1,033½	297	1,780
Lot 3—		Barley.		
First Period	151	54	111	316
Second Period	132	132	60	324
Third Period	87	261	30	378
Fourth Period		361	66	427
Total	370	808	267	1,445
Lot 4—				
First Period	185	59	130	374
Second Period	134	134	52	320
Third Period	139	255	30	424
Fourth Period		378	85	463
Total	458	826	297	1,581

It will be observed that the total consumption of food was greatest with the pigs which were given corn rather than barley. While this was true from the commencement of the feeding period, it was more remarkably so during the later periods of the feeding. This would seem to bear out the idea which to some extent prevails, that barley tends to decrease the appetite, when fed freely to swine for a long period. The difference in the amounts of green food consumed during the different periods is also very noticeable. This was owing largely to the kind of the food fed, and to the succulence of the same. At the first, succulent corn was fed. Of this the consumption was considerable. Then the corn became less succulent, and there was a decrease in the consumption. Toward the last, rape and cabbage were fed. Pigs are fond of both, and both are excellent adjuncts to a heavy grain diet. But the relatively small consumption of green food to the total consumption is also noticeable. The total of grain consumed was 5,339 pounds, and the total of green food 1,183 pounds, that is to say, the consumption of the latter was only a little more than 22 per cent. of the former.

Table XVI. gives the average amount of each food factor consumed per day by the animals in each lot, and the total average daily consumption of these taken singly and together.

TABLE XVI.—Daily Consumption of Food.

	Oats.	Corn.	Barley.	Green Food.	Total.
	Lbs.	Lbs.	Lbs.	Lbs.	Lbs.
Lot 1......	1.23	2.92	....	.96	5.11
Lot 2......	1.34	3.08	....	.88	5.30
Lot 3......	1.10	....	2.40	.79	4.29
Lot 4......	1.36	....	2.46	.88	4.70
Average.	1.26	3.00	2.43	.88	4.85

Reference has already been made to the greater consumption of food by the lots which received the corn as compared with those which received the barley. In this table the contrast is rendered more apparent. The difference in the average consumption of corn per day as compared with barley was a little more than half a pound. But the whole of this is not to be attributed to the diet, as the pigs in lot 3 were low in consumption of food relatively, to some extent at least, from constitutional bias. They were lowest, as will be seen later, in point of gain. The pigs of the first cross consumed somewhat less food per day than those of the second cross. The average daily consumption with the former was 4.70 pounds, and with the latter 5.00 pounds.

Weights of the Animals.—Table XVII. gives the average weight of the pigs in each lot at the commencement of the experiment and at its close, with the total gain made by the pigs in each lot.

TABLE XVII.—Weights and Increase.

	Weight when the Experiment began.	Weight when the Experiment closed.	Total Increase.
	Lbs.	Lbs.	Lbs.
Lot 1	247	601	354
Lot 2	275	603	328
Lot 3	247	526	279
Lot 4	279	567	288
Total	1048	2297	1249

The evenness in the weights of the animals of each class at the commencement of the experiment is marked. Those in lots 1 and 3 of the first cross weighted collectively exactly the same. Those in lots 2 and 4 also weighed practically the same, as the difference in the aggregate weights is only 4 pounds. The total weight of the 6 pigs of the second cross was, however, 60 pounds more than that of the 6 pigs of the first cross, though exactly the same age and fed similarly. But at the end of the experiment this advantage was slightly modified. The 6 pigs of the second cross weighed together at that time but 43 pounds more than the 6 pigs of the first cross. In other words, the latter made more gain by 17 pounds than the former.

Table XVIII. gives the increase in weight made by the pigs in the several lots during the different periods of the experiment.

TABLE XVIII.—Increase During Each Period.

	1st Period.	2d Period.	3d Period.	4th Period.	Total.
	Lbs.	Lbs.	Lbs.	Lbs.	Lbs.
Lot 1	39½	80	101½	133	354
Lot 2	41	72½	104	111	328
Lot 3	57½	44½	103½	73½	278
Lot 4	33½	64	87½	103	288
Average	43	65	99	105	312

Three facts at least revealed by the above table should not escape notice. First, the increase made by the pigs in each of the lots was relatively small during the first period of the feeding, when 75 per cent. of the grain food was composed of oats. The small gains made during this period and the indifferent gains made during the next period, seriously cut down the daily average of the gains. Second, the gains increase as the proportionate quantities of the corn and barley fed increase, with the advancement of the experiment. The warm weather in the summer may have retarded the gains during the first and second periods, but making due allowance for this, the proportion of the oats fed would seem to be too large, and third, the 6 pigs to which corn were fed made together 115 pounds more increase than those to which barley were fed.

A statement of the daily average increase made, in several of its phases is now given:

	Lbs.
Average daily increase of the pigs in Lot 1	1.05
Average daily increase of the pigs in Lot 2	.98
Average daily increase of the pigs in Lot 3	.83
Average daily increase of the pigs in Lot 4	.86
Average daily increase of the pigs in all the Lots	.93
Average daily increase of the pigs of the first cross	.94
Average daily increase of the pigs of the second cross	.92

Average daily increase of the pigs to which corn was fed .... 1.02 Lbs.

Average daily increase of the pigs to which barley was fed .... .84

The average is not high with the pigs in any of the lots, and the reason has been given.

Table XIX. gives the cost of each food factor fed to the animals of the respective lots, and also the cost of these taken together.

TABLE XIX.—Cost of Food Consumed.

	Oats.	Corn.	Barley.	Green Food.	Total.
Lot 1	$2.13	$3.59	....	$0.12	$5.84
Lot 2	2.32	3.78	....	.11	6.21
Lot 3	1.91	....	$3.11	.10	5.12
Lot 4	2.36	....	3.18	.11	5.65
Total	$8.72	$7.37	$6.29	$0.44	$22.82

Oats were the most expensive, relatively, of all the foods fed, as will be apparent by referring to the estimated cost of the food. And so it has been in all our experimental feeding in the west. Because of this, however, we must not leap to the conclusion that they should not be fed at all in fattening animals.

This has not been proved. But until relative values change, they should certainly be used in such feeding with moderation. The lower cost of the barley ration gives only one phase of the story, as will be shown below.

Values and Profit.—Table XX. gives the value of the pigs in each lot when the experiment began and ended, the cost of the food fed, and the resulting profit.

The estimated value put upon the pigs when the experiment began was $3.00 per 100 pounds. This may seem low but pigs were selling very cheaply throughout the entire year. Nor was it so low relatively as the selling price, which was only $3.15 per 100 pounds. The profit on the 6 corn fed pigs was $10.21, while that on the 6 barley fed

TABLE XX.—Values and Profit.

	Lot 1.	Lot 2.	Lot 3.	Lot 4.	Total.
Value on July 13th, when the experiment began....	7.41	$8.25	$7.41	$8.37	$31.44
Cost of food............	5.84	6.21	5.12	5.65	22.82
Total cost...............	$13.25	$14.46	$12.53	$14.02	$54.26
Value on Nov. 2d, when the experiment closed..........	18.93	18.99	16.57	17.86	72.35
Profit......................	$5.68	$4.53	$4.04	$3.94	$18.09

pigs was $7.88. The difference in favor of the former was $2.33. The profit on the 6 pigs of the first cross was $9.72 and on the 6 pigs of the second cross $8.37. The difference in profit in favor of the former was $1.35. Too much, however, must not be made of this fact, as the value of the 6 pigs of the second cross when the experiment closed was $36.85, while that of the 6 pigs of the first cross was $35.50. The value of the former, therefore, exceeded the value of the latter by $1.35. Notwithstanding that the pigs were precisely the same age, and that in all respects they had been similarly treated and fed, the profit made from feeding each animal for 112 days was $1.51.

Cost of Increase.—The cost of making 100 pounds of increase was as stated below:

With the animals in lot 1	$1.65
With the animals in lot 2	1.89
With the animals in lot 3	1.83
With the animals in lot 4	1.96
With the animals that were fed corn...........................	1.77
With the animals that were fed barley.......................	1.90
With the animals of the first cross.............................	1.74
With the animals of the second cross........................	1.93
With the animals in all the lots.................................	1.83

It would seem to be well nigh impossible to make pork at an average cost of $1.83 per 100 pounds, and yet such was the outcome in an experiment which gave relative gains that were only moderately good. Such a result emphasizes the extent of the advantage which the relatively low prices of food in the West give to the Western farmer who is also a grower of pork.

Disposal of the Swine.—At the close of the experiment the pigs were sold to the supply department of the School of Agriculture, and as stated previously, the price obtained was $3.15 per 100 pounds. Prices had touched bottom at the time, hence though the sum received for them may seem very low, it was all that could have been obtained for them in the open market of the Twin Cities of St. Paul and Minneapolis. They were slaughtered by Mr. A. Boss in charge of the department which gives instruction to the students in dressing meats. He pronounced the quality of the meat as super-excellent. The proportion of the side meat was considerably larger than is usually found in Western pigs, the proportion of the lean to the fat was also greater, and the blending of the fat and lean was more perfect. It also made very pleasant and tender eating as may be attested by many competent witnesses in the locality. The proportion of lean in the pigs to which barley was fed was somewhat greater than in those to which corn was fed.

Financial Statement.—

Price received for the pigs on November 2nd, 1896, when the experiment closed		$72.35
Estimated value of the pigs on July 13th, 1896, when the experiment began	$31.44	
Cost of food	22.82	
Total outlay		54.26
Total net profit		$18.09
Net profit on one animal		1.51

Observations.—1. It will be noticed that the food was estimated at average market values in the State. Ordinarily these would be in excess of the cost of production and there-

IMPORTANT FACTS SUMMARIZED.

fore in excess of the home values. In this instance however the difference could not be much because of the very low cost of the food.

2. The value of the manure is supposed to offset the cost of labor, the cost of litter, and the interest on the money invested.

IMPORTANT FACTS SUMMARIZED.

Values—

1. Estimated value per 100 pounds when the experiment began, July 13th, 1896... $3.00
2. Value per 100 pounds when the experiment closed, November 2d, 1896, on the basis of actual receipts 3.15
2. Advance in value per 100 pounds........................... .15

Weights— Lbs.

1. Average weight of the 6 pigs of the first cross when the experiment began... 82
2. Average weight of the 6 pigs of the second cross when the experiment began... 92
3. Average weight of all the pigs when the experiment began... 87
4. Average weight of the 6 pigs of the first cross when the experiment closed... 188
5. Average weight of the 6 pigs of the second cross when the experiment closed... 195
6. Average weight of all the pigs when the experiment closed... 192

Increase in Weight—

1. Average daily increase in weight of the 6 pigs of the first cross.. .94
2. Average daily increase in weight of the 6 pigs of the second cross.. .92
3. Average daily increase in weight of the 6 pigs which were fed corn... 1.02

FEEDING PIGS OF DIFFERENT GRADES.

		Lbs.
4.	Average daily increase in weight of the 6 pigs which were fed barley	.84
5.	Average daily increase in weight of the pigs in all the lots	.94

Food Consumed—

1. Average daily consumption of food by each of the 6 pigs of the first cross.. 4.70
2. Average daily consumption of food by each of the 6 pigs of the second cross... 5.00
3. Average daily consumption of food by each of the 6 pigs to which corn was fed... 5.21
4. Average daily consumption of food by each of the 6 pigs to which barley was fed...................................... 4.50
5. Average daily consumption of food by each pig in all the lots... 4.85

Cost of Increase—

1. Average cost of making 100 pounds of increase with the 6 pigs of the first cross............................... $1.74
2. Average cost of making 100 pounds of increase with the 6 pigs of the second cross......................... 1.93
3. Average cost of making 100 pounds of increase with the 6 pigs to which corn was fed................... 1.77
4. Average cost of making 100 pounds of increase with the 6 pigs to which barley was fed................ 1.90
5. Average cost of making 100 pounds of increase with the 12 pigs in the experiment 1.83

Increase in Value—

1. Average value of each of the 6 pigs of the first cross when the experiment began............................ 2.47
2. Average value of each of the 6 pigs of the first cross when the experiment closed............................ 5.92
3. Average increase in value of each pig of the first cross.. 3.45

4. Average value of each of the 6 pigs of the second cross when the experiment began............................ $2.77
5. Average value of each of the 6 pigs of the second cross when the experiment closed............................ 6.14
6. Average increase in value of each of the 6 pigs of the second cross... 3.37

Profits—

1. Aggregate net profit from feeding 12 pigs for 112 days.. 18.09
2. Average net profit from feeding one pig for the same period.. 1.51

CONCLUSIONS.

The following stand prominent among the conclusions that may be drawn from the experiment:

1. That because of the low relative gains made by the pigs in all the lots during the first period of feeding, and because of the relative increase in the gains subsequently as the proportion of the oats fed was lessened, the conclusion would seem to be legitimate that a diet in which oats is a predominant factor is not the most suitable one that can be fed to pigs while being grown and fattened.

2. That because of the low consumption of food by the pigs to which barley was fed, the conclusion would seem to be legitimate that the free use of barley long continued in growing and fattening pigs tends to weaken the appetite at least to some extent.

3. That because of the low gains made by the pigs to which barley was fed, the conclusion would seem to be fair that a barley diet long continued is not quite so well fitted to make increase in weight as a corn diet, the other adjuncts being the same as in the experiment.

4. That since the pigs of the first cross made somewhat better gains and on less food than those of the second cross this experiment favors the view that they were a little more easily kept.

5. That because of the smallness of the difference in the relative gains of the animals of the two respective crosses, the experiment does not prove that one cross, as such, has any superiority over the other in capacity to make gains.

6. That with the prices of food and meat as in the experiment 100 pounds of pork may be produced at a cost of $1.83 and yielding a profit of $1.32.

www.ingramcontent.com/pod-product-compliance
Lightning Source LLC
Chambersburg PA
CBHW060004230526
45472CB00008B/1943